Paulo Henrique A. da Hora
Sibele B.C. Pergher
Ademir O. Silva

Synthesis, characterization and application of Mg-Al, Zn-Al and Mg-Fe HDLs

Paulo Henrique A. da Hora
Sibele B.C. Pergher
Ademir O. Silva

Synthesis, characterization and application of Mg-Al, Zn-Al and Mg-Fe HDLs

Lamellar Double Hydroxides, Synthesis and Characterization, Memory Effect, Adsorption

ScienciaScripts

Imprint

Cover image: www.ingimage.com

This book is a translation from the original published under ISBN 978-613-9-68987-3.

Publisher:
Sciencia Scripts
is a trademark of
Dodo Books Indian Ocean Ltd. and OmniScriptum S.R.L publishing group

120 High Road, East Finchley, London, N2 9ED, United Kingdom
Str. Armeneasca 28/1, office 1, Chisinau MD-2012, Republic of Moldova, Europe
Printed at: see last page
ISBN: 978-620-8-24210-7

SUMMARY

CHAPTER 1: DETERMINATION OF CRYSTAL PARAMETERS IN MG-AL, MG-FE AND ZN-AL LAMELLAR DOUBLE HYDROXIDES THROUGH X-RAY DIFFRACTION DATA

1- INTRODUCTION

The X-ray diffraction technique is extremely important when it comes to determining crystalline structures, as it offers numerous advantages when it comes to characterizing new materials. Although synthetic HDLs do not show crystallinity comparable to natural materials, parameters such as crystallite size, interplanar distance, available interlamellar space, planar density, volumetric density and crystal lattice parameters can be properly elucidated using the technique (Alvarenga, 2006; Forano et. al, 2006; Albers, 2002). These parameters, explained in this chapter, are extremely important for understanding certain properties of lamellar double hydroxides with a view to modifying them to suit a particular purpose, whether it be doping the crystal for use in electronic and electrochemical applications, obtaining composites with specific characteristics, support for metal catalysts, etc.

X-ray diffraction is relevant and analytically important when it is applied to the study of the crystalline substance that produces the diffraction (Ewing, 2004). Among the various techniques for characterizing materials, X-ray diffraction is the most suitable for determining the crystalline phases present in materials. This is possible because in most crystals, the atoms are arranged in crystalline planes at distances of the same order of magnitude as the wavelengths of the X-rays (Forano et. al, 2006). Different chemical substances can never be expected to develop crystals in which the distance between the planes is identical in all comparable directions, so a complete study of X-rays needs to give a unique result for each substance. When an X-ray beam is incident on a crystal, it interacts with the atoms present, causing the phenomenon of diffraction. X-ray diffraction occurs according to Bragg's Law, which establishes the analogy between the angle of diffraction and the distance between the planes that caused it (peculiar to each crystalline phase) (Ewing, 2004; Jekins & Snyder, 1996). Equation 1 illustrates what is known as Bragg's Law.

$$n\,\lambda = 2d\,.\text{sen}\,(\theta) \qquad \text{[Eq.1]}$$

Bragg's Law states that the incident beams are in phase and collimated. In practice, there is a range around the Bragg angle where the beam hits the crystal. Then there is also a range in which the intensity has a certain appreciable magnitude. This intensity is maximum at the central position of the diffraction peak and halves at the point called "full with at half maximum" (FWHM), i.e. the FWHM is the appreciated measure of the width of a diffraction peak at the point where the intensity halves (Alvarenga, 2006; Ewing, 2004; Jekins & Snyder, 1996; Albers, 2002).

The diffraction peaks for synthetic Double Lamellar Hydroxides (HDLs) are generally refined as belonging to the R-3m space group whose symmetry is rhombohedral and the respective lattice parameters and basal spacing are calculated according to Equation 2 (Forano et. al, 2006), where d is the distance between the planes; h, k and l are the Miller indices; a and c are unit cell parameters:

$$1/d^2 = [4(h^2+hk+k^2)/3a^2] + l^2/c^2 \qquad \text{[Eq.2]}$$

The reference values (Allmann & Jepsen, 1996) associated with the angles, intensity of the diffraction peaks and the respective diffraction planes for hydrotalcite, the most common HDL, are shown in Table 1.

Table 1 - Reference XRD data for Natural HDL Hydrotalcite (Mg-Al-CO_3).

Adapted from Allmann & Jepsen (1996).

2- Theta	Intensity	Basal spacing (d)	Diffraction plane (hkl)
11,64	100,00	7,60333	0 0 3
23,40	34,54	3,8017	0 0 6
34,13	2,25	2,6272	1 0 1
34,82	30,60	2,5765	0 1 2
35,42	1,89	2,5344	0 0 9
39,38	25,32	2,2881	0 1 5
46,85	30,73	1,9391	0 1 8
53,01	7,34	1,7273	1 0 10
56,38	4,76	1,6319	0 1 11
60,64	9,67	1,5270	1 1 0
61,99	9,24	1,4971	1 1 3
63,64	4,12	1,4621	1 0 13
65,92	3,00	1,4170	1 1 6
67,52	1,49	1,3872	0 1 14
71,87	1,18	1,3136	2 0 2
74,74	2,66	1,2701	2 0 5
75,80	3,41	1,2549	1 0 16
79,97	2,37	1,1997	2 0 8
87,48	1,12	1,1150	2 0 11

2 - EXPERIMENTAL PROCEDURE

2.1 - Synthesis of Lamellar Double Hydroxides

Samples of anionic clays from three systems M -M -A^{2+3+-y} , whose ratio, M^{2+} / M^{3+} =2:1, were synthesized using the coprecipitation method at varying pH, which consists of adding a saline solution containing the two cations to be introduced into the lamelas in molar equivalence of value 2:1, to another solution containing the anion to be intercalated (in this case carbonate) in a 2.0 mol.L^{-1} of NaOH. The mixture was subjected to a 24-hour hydrothermal bath (Daute et. al, 2002), then the materials were matured for 72 hours and subsequently filtered and the crystals washed to pH=7.0. The HDL systems synthesized were Mg-Al-CO3, Mg-Fe-CO3 and Zn-Al-CO3.

2.2 - X-ray Diffraction Data Processing

The XRD analyses were processed using a Shimadzu diffractometer, with step size = 0.05 and 2ϴ in the 1.5°-60° range. The X-ray data obtained was processed and the desired parameters for the elucidation of this study were calculated using the H'pert HighScore Plus® software owned by Panalytical©.

3 - DESCRIPTION OF THE PARAMETERS OBTAINED AND RESULTS

Figure 1 shows the X-ray diffractograms for the synthesized materials. The profiles of the diffraction peaks for all the samples worked on in this study show that the materials obtained have good crystallinity. This conclusion is attributed to the width of the diffraction peaks and the profile of the diffractograms shown in Figure 3.1, as can be seen in the illustration.

However, the diffraction intensity values show that few crystals were obtained for the Mg-Fe HDL, as illustrated in Figure 2. The Zn-Al HDL showed a very high diffraction intensity, due to the fact that a large number of smaller crystals were formed. Figure 2 shows the diffraction intensity for the materials used in this study.

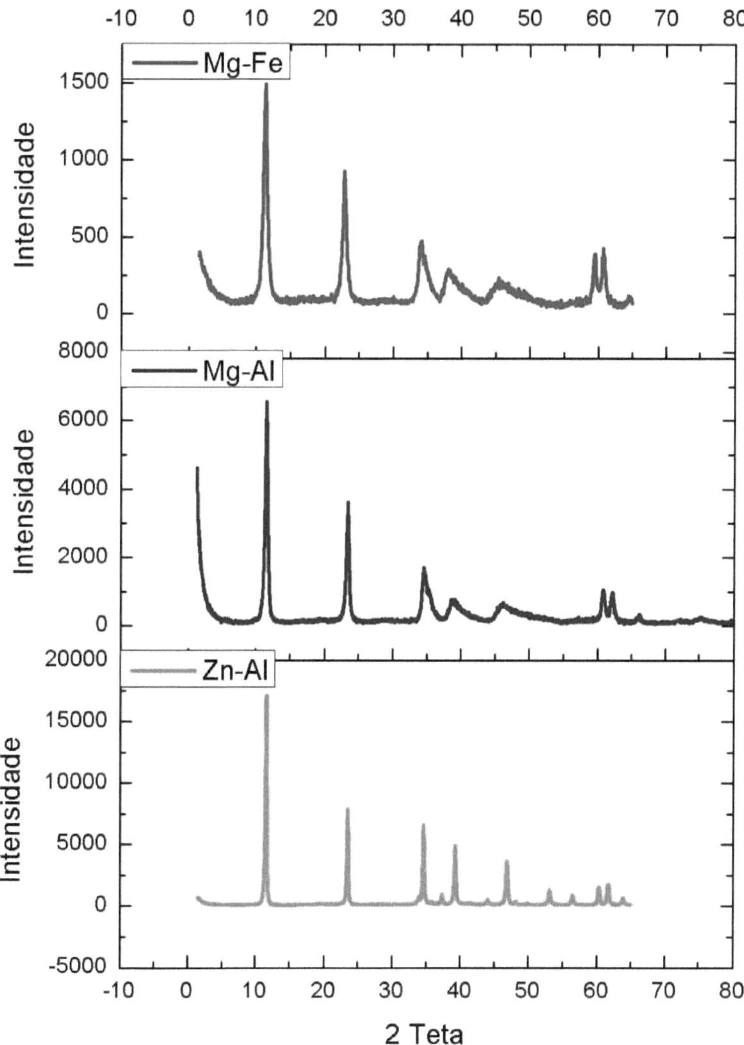

Figure 1. - Diffractograms of the samples synthesized in this study

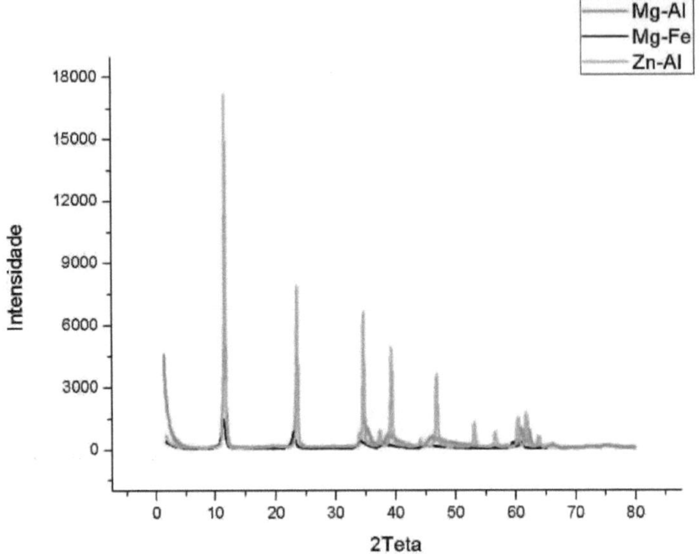

Figure 2- Relationship between the diffraction intensities of the HDLs studied in this work.

Scanning Electron Microscopy (SEM) images confirm the deductions made in Figures 1 and 2, i.e. the fact that the Mg-Fe HDL crystals obtained had good crystallinity, but few crystals were obtained. Also, the fact that many Zn-Al HDL crystals were obtained, which resulted in high diffraction intensity. Figure 3 illustrates the scanning electron microscopy images for the materials synthesized in this study: (3a) Mg-Al, (3b) Zn-Al and (3c) Mg-Fe.

Figure 3 - Scanning Electron Microscopy images for HDLs: Mg-Fe (a), MgAl (b) and Zn-Al (c)

(a)

(b)

(c)

3.1 - Volumetric Density

As with the calculation of the density of any material, it illustrates the mass of a given specimen or group of atoms per volume of material. Based on the atomic model of rigid spheres for the unit cell of the crystal structure of any material, the volumetric density (ρ_v) is obtained from Equation 3 (Callister, 2002), where n is the number of atoms per unit cell, A is the atomic/molecular mass of the material, V is the volume of the unit cell of the material and N is the Avograd number ($N=6.02.10^{23}$).

$$\rho_v = \frac{n.A}{V.N} \qquad \text{[Eq. 3]}$$

However, for the purposes of this study, only the volumetric density of the metals that make up the brucite-like layers of the HDLs will be taken into account.

Taking into account the fact that a unit cell is a specific repetitive unit of the crystal that simply reflects the structure of the whole material, as shown in Figure 4, in HDLs whose ratio M^{2+}/M^{3+} is equal to 2:1 the amount of metals in a unit cell is equal to $6M^{2+}$ to $3M^{3+}$.

Figure 4 - Distribution of cations per unit cell for HDLs of different compositions.

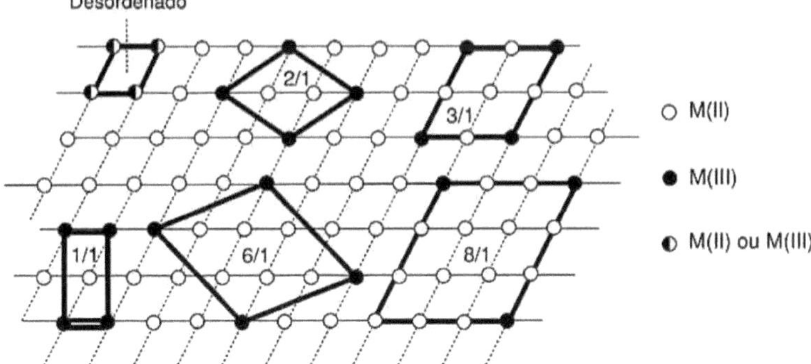

Although these were not metal specimens, but their respective cations, the atomic mass of the metal was taken into account, since the loss of electrons it exerts is negligible in relation to the effect it has on the atomic mass (Atkins & Jones, 2009). Therefore, the following atomic masses were considered, as shown in Table 2.

Table 2- Atomic masses of the metal specimens used in the respective studies.

Adapted from Callister (2002).

Specimen	Atomic Mass (one)
Mg	24,31
Al	26,98
Zn	65,39
Fe	55,85

Certainly, as can be deduced from Equation 3, larger metals tend to occupy a greater volume, so for a given volume, we will have a greater volumetric density of larger specimens per unit cell. However, it is necessary to take into account the influence exerted by the volume of the unit cell of each material, since this quantity takes into account the atomic radius of the specimens involved.

In the case of HDLs, which have a hexagonal system, the volume of the unit cell is calculated using the relationship expressed in Equation 4, where Vc = volume of the unit cell and r is the atomic radius of the specimens involved.

$$Vc = 24r^3\sqrt{2} \qquad \text{[Eq. 4]}$$

Table 3 illustrates the unit cell volume values for the HDLs used in this study.

Table 3 - Unit cell volume for the synthesized HDLs

HDL	Unit Cell Volume ($Å^3$)
Mg-Fe-CO3	131,0
Mg-Al-CO3	127,5
Zn-Al-CO3	123,6

The volumetric density data for each metal in the unit cell of an HDL is shown in Table 4. As expected, specimens with a higher atomic mass have a higher volumetric density per unit cell volume.

Table 4 - Volumetric Density of the Constituent Metals of the Synthesized HDLs.

HDL	Volumetric density of M^{2+} for each $Å^3$	Volumetric density of M^{3+} for each $Å^3$
Zn-Al	$5,273.10^{-22}$	$1,100.10^{-22}$
Mg-Al	$1,900.10^{-22}$	$1,050.10^{-22}$
Mg-Fe	$1,850.10^{-22}$	$2,100.10^{-22}$

The theoretical provisions regarding the volume density of the constituents of the HDL layers show that the Zn-Al material has a greater possibility of exchange in the crystal lattice when it comes to divalent specimens, and is therefore more likely to be doped for electronic, magnetic and electrochemical applications. In the case of trivalent specimens, the theoretical possibility of greater doping success is in Mg-Fe HDL. However, as these are theoretical provisions, It Is up to the synthesis result to determine the origin of these attributes, and it is necessary to obtain a relevant crystalline fraction

3.2 - Planar Atomic Density

This refers to the number of atoms per unit area of a given crystallographic plane. Generally, it is more common to use as a reference for determining this quantity the plane that intersects the entire structure of the unit cell, as this divides it in half. Equation 4 is used to calculate the planar atomic density (Callister, 2002).

$$\rho = \frac{n\ of\ atoms\ whose\ centers\ are\ intercepted\ by\ the\ selected\ area}{selected\ area} \qquad [Eq. 4]$$

The area of the plane of interest was calculated by multiplying the crystalline parameter a by the parameter c (Equations 6 and 7, respectively). Taking into account the fact that lamellar double hydroxides belong to the hexagonal crystalline system, the number of atoms per unit cell is equal to 4 (Callister, 2002).

Table 5 shows the data for calculating the planar density. As the Zn-Al system has a shorter distance between neighboring atoms and a greater unit cell height, it therefore has a greater number of atoms per unit cell area.

Table 5 - Data used to obtain the Planar Density

HDL	Parameter a (À)	Parameter c (À)	a . c (At)^2	Planar Density (atoms per square angstrom)
Zn-Al	3,066	22,701	69,6013	0,057
Mg-Al	3,54	26,76	94,7304	0,042
Mg-Fe	3,10	23,22	71,982	0,056

These illustrated data are confirmatory of those obtained in the atomic density calculation, since these parameters are closely related to the radius of the specimens involved. In theory, therefore, the Zn-Al and Mg-Fe materials present promising possibilities for exchanging cations in the unit cell.

3.3 - Crystal size

The Scherrer equation illustrated in Equation 5 (Alvarenga, 2006) is used to calculate the crystallite size (ε_ω) of any material. Where K is the form factor (a constant with a value of 0.9 for analysis with radiation from Cu), λ is the X-ray wavelength, β is the "full with at half maximum" (FWHM), i.e. the FWHM is the appreciated measure of the width of a diffraction peak at the point where the intensity halves and ϴ is the diffraction angle.

$$\varepsilon_{hkl} = \frac{K.\lambda}{\beta.cos\theta} \qquad \text{[Eq.5]}$$

This equation is widely used to evaluate the grain size of materials and gives good results when used as a first approximation for calculating the average size of a crystallite.

To determine the crystallite size of lamellar double hydroxides, it is necessary to consider K=0.9 (Forano et. al, 2006). Explaining the discrepancies between crystallite sizes is extremely complex, since several factors can influence them, such as nucleation, the method of synthesis, the nature of the metal cations, the purity of the reagents, etc. Essentially, because in this study, all the materials were synthesized under the same conditions. Table 6 shows the results obtained for crystallite size using the Scherrer equation (Equation 5). Although the Mg-Fe HDL had a smaller crystallite size, the SEM data shows that few crystals were obtained during the synthesis of this material. The Mg-Al and Zn-Al materials, on the other hand, had larger crystallite sizes, but higher yields during the synthesis process. Compared to the literature, the values are relevant, as these data show crystallites in the range of approximately 90nm.

Table 6 - Crystallite size values using Scherrer's equation.

Sample	$\varepsilon_{h\,l\,k}$ (nm)
Mg-Al-CO3	177
Zn-Al-CO3	57
Mg-Fe-CO3	28

3.4 - Crystal lattice parameters

Synthetic HDLs have a hexagonal crystalline system, so $a=b\neq c$.

The network parameter a is calculated using the expression illustrated in Equation 6 (Rodrigues, 2007). The parameter a is extremely important in the case of HDLs, as it perfectly reproduces the distance between canons in neighboring octahedrons in brucite-type layers, since it takes into account the interplanar distance of the plane that intersects the brucite-type layer.

$$a = 2 \times d_{(110)} \quad \text{[Eq.6]}$$

The parameter c is calculated using the expression illustrated in Equation 7. The importance attributed to this parameter is that it reproduces the height of the unit cell of the respective HDL.

$$c = 3 \times d_{(003)} \quad \text{[Eq.7]}$$

Table 7 shows the unit cell heights of the materials synthesized in this study (Equation 7). These data illustrate that the Mg-Al HDL showed an anomaly in this respect, since, compared to the other materials, it has a much higher unit cell value. The values for the Zn-Al and Mg-Fe HDLs are within the expected range based on the specialized literature.

Table 7 - Unit cell height of the synthesized HDLs

Sample	Parameter c (A)
Mg-Al-CO3	26,760
Zn-Al-CO3	22,701
Mg-Fe-CO3	23,220

The data for parameter a (Equation 6), which indicates the distance between the cations in the brucite-type octahedrons, illustrates the trend related to the ionic radii of the respective ions used in the synthesis, i.e. for specimens with a larger ionic radius, a greater distance between atoms is observed in the octahedrons of the brucite-type layers. The reciprocal is also true. Table 8 illustrates the data obtained.

Table 8 - Distance between the cations in the brucite-type octahedrons in the HDLs synthesized in this work.

Sample	Distance (A)
Mg-Al-	3,540
CO3 Zn-Al-CO3	3,066
Mg-Fe- CO3	3,100

3.5 - Interplanar Distance Considering the Plane <hkl> with the Highest Intensity of Diffraction

The interplanar distance considering the plane with the highest diffraction intensity, in the case of lamellar double hydroxides the <003> plane, reflects the contribution of the height of the brucite-type layer of an HDL added to its respective interlamellar space.

The calculation of this crystalline parameter is carried out following Bragg's Law, as illustrated above in Equation 1 (Rodrigues, 2007).

When determining this parameter, it is necessary to consider the peak with the highest diffraction intensity. In the case of lamellar double hydroxides, this concerns the (003) plane, where the value of d in this plane represents the sum of the lamella thickness and the height of the interlamellar region. The results obtained for the Mg-Fe and Zn-Al HDLs are entirely consistent with the data in the literature, which report interlamellar distance values for HDLs in the range of 7.6 to 7.8 Â (Cavani et al., 1991; Santos & Corrêa, 2011). Table 9 illustrates the d_{003} values. The data for the Mg-Al HDL is totally unusual.

Table 9 - d_{003} values of the HDLs synthesized in this work.

Sample	d_{003} (A)
Mg-Al- CO3	8,920
Zn-Al-CO3	7,567
Mg-Fe- CO3	7,740

3.6 - Interlamellar Space Available

Using X-ray diffraction information, it is possible to estimate the available interlamellar space (dl⅛re), using the expression illustrated in Equation 8 (Beaudot et. al, 2004; Bonnet et. al, 1996; Ennadi et. al, 1994; Meyn et. al, 1993). al, 1993), where H_{layer} represents the thickness of the layer containing the octahedrons (around 0.20 nm), $_{dobs}$ represents the basal spacing and LH represents the hydrogen bonding between the layers and the interlamellar anions (around 0.27 nm).

$$d_{livre}=d_{obs} + (H_{camada}-2_{LH}) \quad \text{[Eq.3.8]}$$

The values for the free interlamellar space ($_{dlivre}$) are directly linked to the basal spacing and the height of the unit cell, so the Mg-Al HDL had a much higher available interlamellar space than the other materials. This makes it possible to insert larger molecules into this region of the material, making it more versatile in certain applications.

Table 10 - $_{dl\ vire}$ values of the HDLs synthesized in this work.

Sample	$_{free}$ **(nm)**
Mg-Al- $_{CO3}$	55,200
Zn-Al-CO3	41,670
Mg-Fe- $_{CO3}$	43,400

4 - CONCLUSIONS

Although HDL $_{Mg-Fe-CO3}$ is the material that has shown some relevant properties, such as atomic and planar density, and even with a smaller crystallite size than the others, the XRD and SEM data, illustrated in Figures 1 and 2 respectively, show that few crystals were obtained in the synthesis of the material, which makes its potential unfeasible, since a material is needed that can combine desired properties with easy reproduction in the laboratory.

HDL $_{Zn-Al-CO3}$, on the other hand, has promising potential, since the predictions for its atomic and planar density show good prospects, and, in addition, the synthesis yields show that this material is easily obtained in desirable crystallinity. In terms of feasibility for applications in electronics and electrochemistry, this material presents possibilities more suited to those required in such processes, or it is easier to turn it into such a material.

The Mg-Al HDL showed a structural anomaly: its basal spacing ($_{d003}$, calculated using Bragg's law) was extremely high. In this case, the size of the available interlamellar space is also high. Therefore, the possibility of intercalating certain larger specimens makes this material a promising component in obtaining a composite combining the potential of HDL with that of a larger polymer, for example. There is also the possibility of inserting larger pillars or catalysts. Another notable aspect for the application of this material is in the pharmaceutical field. As the interlamellar space is very large, it is possible to obtain drugs with larger quantities of active ingredient, as well as to combine HDLs with active ingredients whose molecules are very large.

REFERENCES

ALBERS, A. P. F.; **Ceràmica**, n. 48, pag. 34, 2002.

ALLMANN, R. & JEPSEN, H. P.; **Neues Jahrbuch Fur Mineralogie, Monatshefte: Die Struktur Des Hydrotalkits**, 1969.

ALVARENGA, D. R. **Semiconductor Ceramics: A Structural Study of Thick Films of CDs**. Master's dissertation. Postgraduate Program in Physics, Federal University of Minas Gerais, 2006.

ATKINS, P.; JONES, L. **Principles of Chemistry: Questioning modern life and the environment**. 6ª Ed. Bookman: São Paulo, 2006.

BEAUDOT, P.; ROY, M. E. DE.; BESSE, J.P.; **Journal of Solid State Chemistry**, n.177, pag. 2691,2004.

BONNET, S.; FORANO, C.; ROY, A. DE; BESSE, J.P.; MAILLARD, P.; MOMENTEAU, M.; **Chem. Mater**., n. 8, pag. 952, 1996.

CALLISTER, W. D. **Ciência e Engenharia de Materiais: Uma Introduçâo**, 5° ed., LTC : Rio de Janeiro, 2002.

DAUTE, P.; FOELL, J.; LANGE, I.; KUEPPER, S; WEDL, P.; KLAMANN, JD. **US Patent 6,362,261**, 2002.

DEL ARCO, M; FERNANDEZ, A.; MARTiN, C.; RIVES, C.; **Applied Clay Science**, n. 36, pag. 133, 2007.

ENNADI, A.; KHALDI, M.; ROY, A. DE; BESSE, J.P.; **Mol. Cryst. Liq. Cryst**., n. 244, pag. 373, 1994.

EWING, G. W. **Instrumental Methods of Chemical Analysis**. Edgard Blücher, São Paulo. 2004.

FORANO, C.; HIBINO, T.; LEROUX, F.; TAVIOT-GUÉHO, C. Layered Double Hydroxides. In: BERGARA, F.; THENG, B.K.G.; LAGALY, G.; **Handbook of Clay Science**; Elsevier: Amsterdam, p. 1021,2006.

JENKINS, R.; SNYDER, R.L. **Introduction To X-Ray Powder Diffractometry**. John Wiley: New York, 1996.

MEYN, M.; BENEKE, K.; LAGALY, G.; **Inog. Chem**., n. 32, pag. 1209, 1993.

RODRIGUES, J.C. **Sintese, Caracterizaçâo e Aplicações de Argilas Anionicas do Tipo Hidrotalcita**, Dissertação De Mestrado, Programa de Pôs-graduaçâo em Quimica, Universidade Federal do Rio Grande do Sul, 2007.

CHAPTER 2: EVOLUTION OF THE "MEMORY EFFECT" IN LAMELLAR DOUBLE HYDROXIDES MG-AL AND ZN-AL EXPOSED TO ATMOSPHERE

1 - INTRODUCTION

Some double hydroxides exhibit the property of regenerating the initial structure of the material after being subjected to thermal decomposition, which is known as the "memory effect" (Rodrigues, 2007). According to the literature, when these materials are calcined at temperatures close to 500°C, the interlamellar anions decompose and the material is almost completely dehydroxylated, thus causing the loss of the lamellar compound (confirmed by X-ray diffraction data, through the absence of diffraction planes 003, 006, 009). The result of this process is a solid solution of a double hydroxide-oxide of M^{2+} and M^{3+} (Reis, 2009), a periclassium-type compound. Figure 1 shows the structural changes that occur during the calcination process.

Figure 1 - Post-calcination structural alteration of a lamellar double hydroxide. Adapted from Reis (2009).

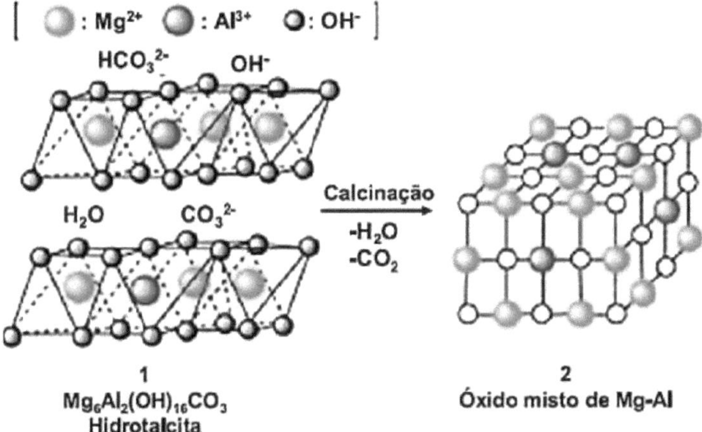

A simple addition of water or exposure to the environment of this formed oxide is supposed to reconstruct the structure of the precursor lamellar double hydroxide. However, the occurrence of the "memory effect" is subject to the proper formation of the mixed oxide (Stanimirova et. al, 2004). At extremely high temperatures (close to 1000°C), a spinel-like compound is formed (Velu et. al, 1999), which is stable under ambient conditions, so regeneration of the laminar structure is not possible at this stage.

This structural regeneration resulting from exposure to the atmosphere is due to the adsorption of moisture and CO_2 from the atmospheric air (Rodrigues, 2007). When added to water, the carbonate anion comes naturally from the dissolution of atmospheric CO_2 in the solvent (Wang, 2007).

The memory effect in HDLs of the Mg-Al system is well documented in the literature, but its occurrence for the Zn-Al system is controversial (Newman & Jones, 1998; Vaccari, 1998).

With this in mind, this work aims to study and evaluate the memory effect of hydrotalcite-type compounds based on Mg-Al and Zn-Al when exposed to the atmosphere.

2 - EXPERIMENTAL

The anionic clay samples from the two systems M -M -A^{2+3+-y} , whose ratio, $M^{2+}{}_{/M3+=2}$:1, were synthesized using the coprecipitation method at variable pH, which consists of adding a saline solution, made up of the two cations to be introduced into the layers of the material in equivalence M^{2+} :M^{3+} = 2:1, to another solution containing the anion to be intercalated between the lamellae (in this case carbonate) in a 2.0 mol.L^{-1} of NaOH. The mixture was subjected to a hydrothermal bath for 24 hours (Daute et. al, 2002), then the materials were matured for 72 hours and subsequently filtered and the crystals washed to pH=7.0. The HDL systems synthesized were Mg-Al-CO3 and Zn-Al-CO3. The material was characterized using XRD analyses, which were processed using a Shimadzu diffractometer, with step size = 0.05 and 2ϴ in the 1.5°-60° range.

The x-ray diffraction data show that the desired materials were successfully obtained, as illustrated in Figure 2.

Figura 2 - Diffractograms of the materials synthesized in this study.

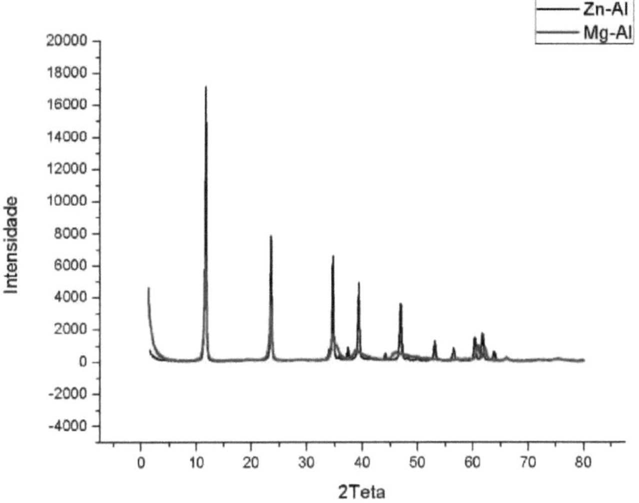

The lamellar double hydroxide samples from the Mg-Al and Zn-Al systems were subjected to the calcination process at a heating rate of 20°C/min up to a temperature of 470°C. After calcination, the materials were subjected to the action of the local atmosphere for up to 2 weeks. X-ray diffraction analyses were carried out during certain periods of time in order to verify the action of the atmosphere on the restructuring of the original organization of the precursor material. The samples were analyzed after the end of calcination, 1 week of exposure to the atmosphere and 2 weeks of exposure. In addition, the X-ray diffractogram of the sample with the longest exposure time (2 weeks) was compared with the original sample.

3 - RESULTS

The dynamics of structural reorganization ("memory effect") when the calcined Zn-Al and Mg-Al HDLs were exposed to the atmosphere showed similar dynamics. Both samples showed the occurrence of the "memory effect" at the same minimum exposure time interval, 2 weeks. After a week's exposure, the diffractograms of the samples showed profiles similar to those of the calcined samples. However, during this exposure time for restructuring, the material did not acquire crystallinity similar to that of the precursor HDL samples, but it did acquire good crystallinity.

Figure 3 illustrates the structural evolution of calcined Mg-Al HDL during exposure of the sample to the atmosphere over a period of up to 2 weeks. The reference "Mg-Al Calc" was used for the calcined sample. For the sample after 1 week of exposure, the reference "Mg-Al 1Sem" was used. For the sample after 2 weeks of exposure, the reference "Mg-Al 2Sem" was used.

Figura 3 - Evolution of the "Memory Effect" for Calcined Mg-Al HDL

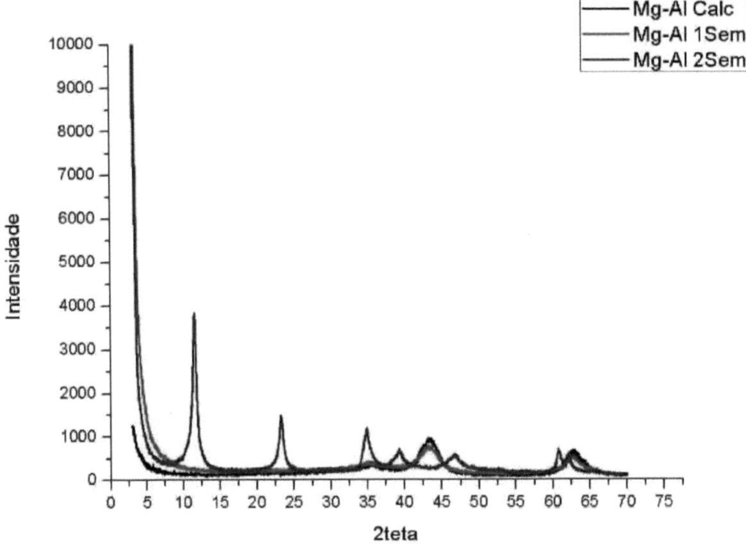

As mentioned above, the maximum exposure time did not allow this material to acquire a structural organization similar to that of the precursor sample. Figure 4 shows a comparison between these samples (Mg-Al 2sem and Mg-Al)

Figura 4 - Comparison between samples of calcined material exposed to the atmosphere for 2 weeks and the HDL precursor

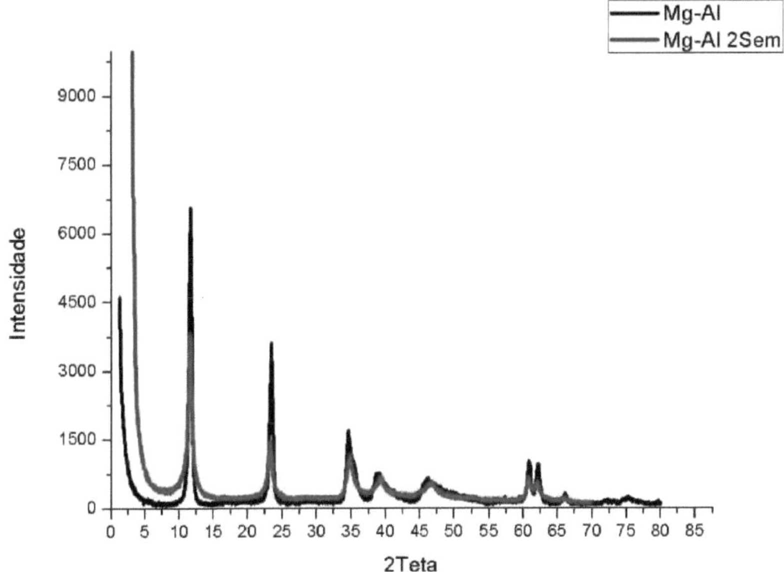

As anticipated, the results with the Zn-Al HDL samples showed the same dynamics as those with Mg-Al. Although they did not show the same crystallinity as the precursor sample, very crystalline phases were obtained.

Figure 5 shows the structural evolution of the calcined Zn-Al HDL during exposure of the sample to the atmosphere over a period of up to 2 weeks. The reference "Zn-Al Calc" was used for the calcined sample. For the sample after 1 week of exposure, the reference "Zn-Al 1Sem" was used. For the sample after 2 weeks of exposure, the reference "Zn-Al 2Sem" was used.

Figure 6 shows a comparison between the samples of calcined material exposed to the atmosphere over a period of 2 weeks and the precursor Zn-Al HDL sample.

Figura 5 - Structural evolution of calcined HDL Zn-Al during exposure of the sample to the atmosphere over a period of up to 2 weeks

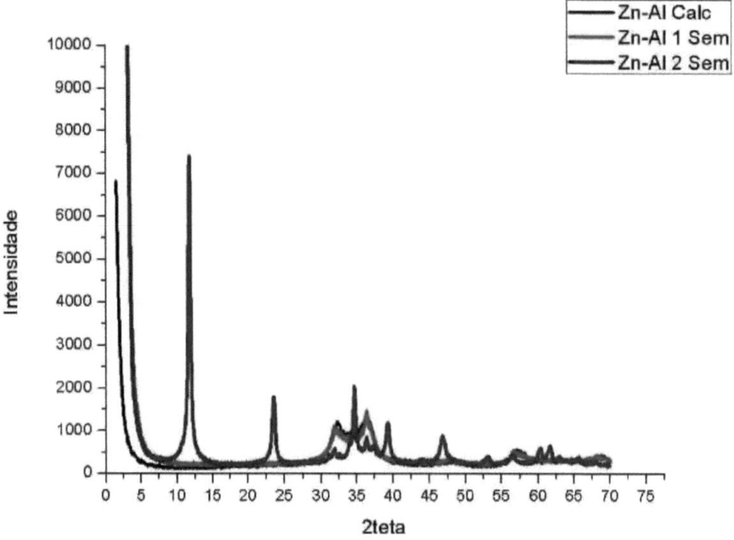

Figura 6 - Comparison between samples of calcined material exposed to the atmosphere for 2 weeks and the precursor Zn-Al HDL sample

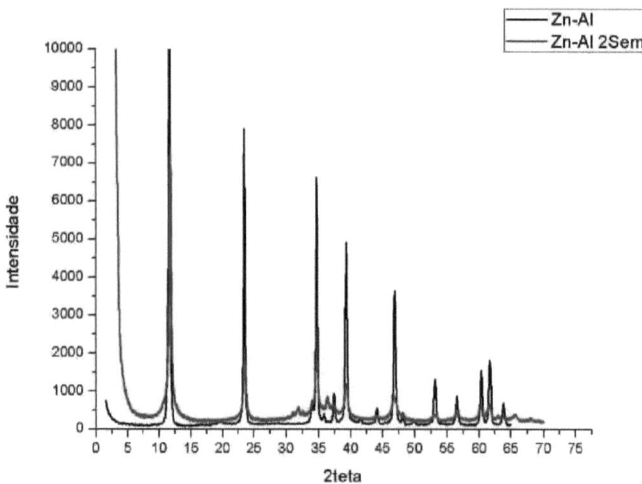

4 - CONCLUSIONS

The data obtained in this study confirms the information reported in the specialized literature regarding the possibility of the so-called "memory effect" occurring in HDLs of the Mg-Al and

Zn-Al systems. However, it was observed that this structural reconstruction is achieved within a minimum of 2 weeks of exposure to the atmosphere.

REFERENCES

NEWMAN, S.P.; JONES, W.; **New Journal of Chemistry**, n. 22, pag. 105, 1998.

REIS, M. J.; **Synthesis and Characterization of Lamellar Double Hydroxides Prepared in the Presence of Organic Polymers or with Intercalated Macromolecules**, PhD Thesis, Postgraduate Program in Chemistry, University of São Paulo, Ribeirão Preto, 2009.

RODRIGUES, J.C. **Sintese, Caracterizaçâo e Aplicações de Argilas Aniônicas do Tipo Hidrotalcita**, Dissertação de Mestrado, Programa de Pôs-Graduaçâo em Quimica, Universidade Federal do Rio Grande do Sul, Porto Alegre, 2007

STANIMIROVA, T.; PIPEROV, N.; PETROVA, N.; KIROV, G.; **Clay Minerals**, n. 39, pag. 177, 2004.

VACCARI, A.; **Catalysis Today**, n. 41, pag. 53, 1998.

VELU, S.; SUZUKI, K.; OKASAKI,M.; OSAKI, T.; TOMURA, S.; OHASHI, F.; **Chem. Mater.**, n. 11, v. 8, pag. 2163, 1999.

WANG, H.; CHEN, J.; CAI, Y.; JI, J.; LIU, L.; TENG, H.H.; **Applied Clay Science**, n. 35, pag. 59, 2007.

CHAPTER 3: KINETIC PARAMETERS OF THERMAL DECOMPOSITION METHODS OF HDLS OBTAINED BY THERMOGRAVIMETRY

1 - INTRODUCTION

Since lamellar double hydroxides have become an extremely promising material due to their range of reaction applications, especially their application in catalysis and electrochemistry, under various experimental conditions, knowledge about the kinetic mechanism of thermal decomposition of this type of material is extremely important, since it makes the planning of certain processes, in which this material is applied in certain temperature ranges, very close to the real thing.

As soon as a lamellar double hydroxide is subjected to a thermal decomposition process, it is observed that during this process metaphases are formed in certain temperature ranges. The literature reports the existence of three metaphases during the thermal decomposition process: HT-D (referring to the stage of thermal decomposition where complete dehydration of the lamellar double hydroxide is achieved), HT-B (attributed to the metaphase existing when the material is totally dehydrated and partially deoxidized) and MO (designated to the phase acquired by the material when, at high temperatures, the formation of a mixed hydroxide-oxide of the constituent metals of the lamellar double hydroxide in question (Stanimirova et.. al, 2004). HT-D metaphase occurs in a temperature range of approximately 150°C. HT-B occurs at a temperature of around 250°C. The MO phase is formed at temperatures above 450°C and is reversible due to the so-called "memory effect" that exists in lamellar double hydroxides. However, if subjected to temperatures above 850°C, a spinel phase is formed (MgO and MgAl2O4) which is stable under natural conditions and irreversible in terms of the probability of returning to the original structure of the precursor HDL (Forano et. al, 2006). Figure 1 illustrates a schematic of the transformations that occur in a lamellar double hydroxide as a result of being subjected to high temperatures and the approximate temperature ranges in which each event takes place.

Figura 1 - Metaphases during the thermal decomposition of a lamellar double hydroxide

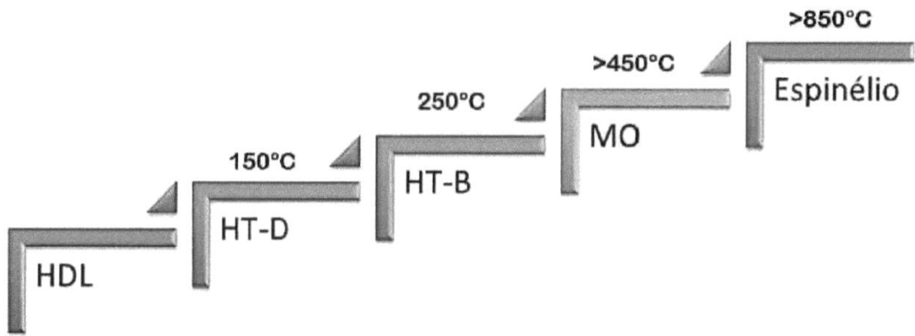

Figure 2 shows a simulation of the structures that these metaphases acquire when they are formed.

Figura 2 - Illustration of the structural arrangement of the metaphases obtained during the thermal decomposition process of an HDL. Adapted from Stantimirova et. al (2004). HT-D: Dehydrated Material; HT-B: Partially Deoxidized Material; MO: Double Oxide.

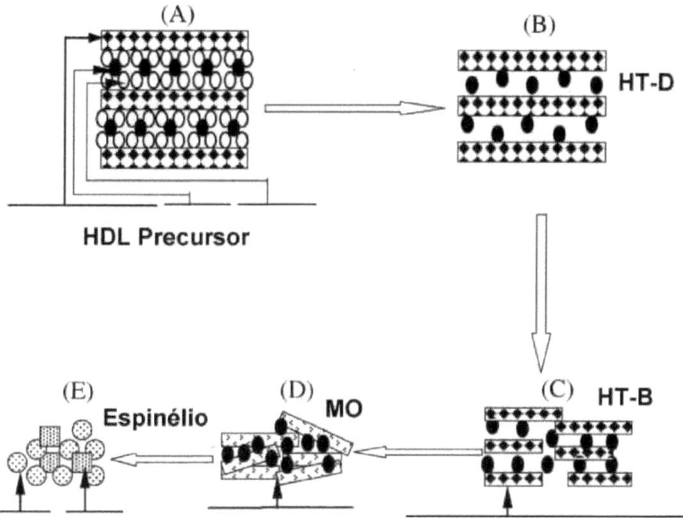

During the metaphase formation process, the literature reports that the dehydration dynamics of an lamellar double hydroxide is a unique process, which progresses as the temperature rises. During this stage, only the loss of intrinsic and extrinsic water is reported, the hydroxyls present in the brucite-like layers do not undergo any changes and the interlamellar anion presents its symmetry and function preserved. Deoxidization is a step that takes place in several similar stages in temperature ranges of 220-600°C. In all, the existence of three deoxydrylation stages is indicated in the literature. It is estimated that at each stage, approximately 1/3 of OH^- is removed from the structure (Valente et. al, 2000).

Based on all the information contained in the literature about the thermal decomposition of lamellar double hydroxides, this study is based on the evolution of the activation energies during the thermal decomposition stages in order to compare this reaction energy behavior with the information contained in the specialized bibliography.

Chemical kinetics is extremely important, since it studies not only the speed of chemical reactions, but also the factors involved in this aspect.

Svante Arrhenius was the great forerunner of studies on the effect of temperature on the kinetics of a chemical reaction in the 19th century. One of Arrhenius' great discoveries was that by graphically expressing the natural logarithm of the reaction rate constant (Ln k) versus the inverse of the temperature at that stage of the reaction (1/T), a linear continuum is obtained. In this context, the intersection of the linear trace with the y-axis (x=0) is called the pre-exponential factor or Arrhenius parameter, represented by ko (Atkins & Jones, 2006) or A (Atkins, 2006). The angular coefficient of this linear sequence, the slope of the line, was considered to be equal to - Ea/R, where Ea is the so-called activation energy and R is the universal gas constant. A and Ea are called reaction kinetic parameters and depend on the order of the reaction being studied. In practical terms, the activation energy refers to a kind of energy step that needs to be overcome in order for the specimens (molecules, ions or groups) involved to carry out the reaction. It is only specimens with sufficient energy to overcome this "step" that make the reaction occur, i.e. the formation of the products of the process. The pre-exponential factor indicates the frequency at which molecular collisions occur (White & Catallo, 2011). Equation 1 illustrates the mathematical expression of the Arrhenius model.

$$K(T) = k_0 e^{-Ea/R} \qquad \text{[Eq. 1]}$$

Since then, various models for describing the kinetics of a chemical reaction have taken the Arrhenius model into account, and methods using thermogravimetric analysis data have been widely used for the kinetic study of processes. In addition, the literature does not consider traditional kinetic study methods to be reliable for applications to complex reactions, so derivative models have been developed (Mackenzie, 1979).

Reaction kinetics using thermochemical methods can be determined in two ways: using static (isothermal) or dynamic (non-isothermal) methods. For both cases, an extremely relevant factor for the study is the calculation of the conversion rate (α), which, as the name suggests, relates the amount of mass existing at a given temperature (m) compared to the initial (m_0) and final (m_f) mass in the process. Equation 2 illustrates the calculation of the conversion rate during pyrolysis.

$$\alpha = (m_0 - m) / (m_0 - m_f) \qquad [Eq. 2]$$

Thus, thermal decomposition of the solid with increasing temperature (dα / dt) can be expressed in terms of the product between two functions, one of which depends on temperature, k(T), and the other on the conversion rate, f(α), as illustrated in Equation 3.

$$d\alpha / dt = k(T).f(\alpha) \qquad [Eq. 3]$$

By substituting the Arrhenius equation, Equation 1, into the expression that reflects the thermal decomposition of a solid with increasing temperature, Equation 4.3, we get the expression shown in Equation 4.

$$d\alpha / dt = k_0 e^{-Ea/R}. f(\alpha) \qquad [Eq. 4]$$

Bearing in mind that these thermal decomposition processes are governed by a programmed heating ratio (β), it is necessary to take this into account. As β= dT / dt, substituting Equation 4 we obtain Equation 5.

$$d\alpha / f(\alpha) = [k_0 / \beta] . e^{-Ea/R} dT \qquad [Eq. 5]$$

However, f(α) depends on the amount of sample and, obviously, the temperature. By integrating Equation 5 from the initial temperature of the thermal decomposition process, T_0, which corresponds to α_0, up to a given temperature, Ttₐₗ ⅛<, where α= $a_{màx}$, we get Equation 6.

$$\int_{\alpha}^{\alpha\ m\acute{a}x} \frac{d\alpha}{f(\alpha)} = k0 \int_{T}^{Tm\acute{a}x} e^{-Ea/RT} dT \qquad [Eq\ 6]$$

As a pyrolysis reaction is quite complex, the form of f(α) in Equation 6 can be quite difficult to deduce, making the process a complicated description of reaction kinetics.

To this end, Biagini et. al (2002) considered that the process obeys a specific reaction order, according to Equation 7, where n is equal to the reaction order considered. Graphically expressing Equation 7 versus time for various reaction orders, the approximate reaction order of a given process is determined by the line of greatest linearity obtained (R^2) (Brown,

2004; Orfâo & Figueiredo, 2001; Osawa, 1965).

$$f(\alpha) = (1-\alpha)^n \qquad \text{[Eq. 7]}$$

Osawa (1965) applied an empirical approximation to the integrals in Equation 4.6, resulting in Equation 8. Graphically, considering the same conversion rate, the angular coefficient of a graphical relationship between log (β) versus 1/T is equal to -0.4567[Ea / R].

$$\text{Log } \beta = -0{,}4567[E_a / RT] + [\log (k_0.E_a / R) - \log f(\alpha) - 2{,}315] \qquad \text{[Eq. 8]}$$

Kissinger (1956), based on the temperature of the maximum rate of mass loss, integrated Equation 6 successively, obtaining Equation 9. Graphically, Ln (β / T_{max}^2) versus $1/T_{max}$, has an angular coefficient equal to - Ea / R.

$$\text{Ln } (\beta / T_{máx}^2) = [\text{Ln } (K_0 R / Ea) - \text{Ln } f(\alpha)] - E_a / RT_{máx} \qquad \text{[Eq. 9]}$$

The "Model Free Kinetics" is based on Vyazovkin's theory (Vyazovkin et. al, 1997; Vyazovkin et. al, 1996) and presents the same mathematical deduction as Kissinger (1956), however, accepting not the temperature of the maximum rate of mass loss, but adopting the respective conversion temperatures, assuming that the activation energy does not remain constant during the reaction. Graphically, by plotting Ln (β / T^2) versus 1/T, we have -Ea / R equal to the angular coefficient of this function.

The Starink method (Starink, 1996) relates the same variables as the other methods, and is considered in the literature to be an adaptation of the Osawa method for reactions in which the variation in activation energy is minimal with the change in the heating ratio, with the calculation of Ea being the angular coefficient of the relationship between Ln [$\beta / T^{1,8}$] as a function of 1/T. Equation 4.10 expresses Starink's model. Where C is a constant from the integration method used in the mathematical deduction and $A=1.007 - 1.2.10^{-5} .E_a$.

$$\text{Ln } [\beta / T^{1,8}] = -A [E_a / RT] + C_1 \qquad \text{[Eq. 10]}$$

The Flynn and Wall method (Flynn & Wall, 1966) uses the least squares method to determine the estimated activation energy (E_a). Graphically, by plotting Ln β versus 1/T (the same function as in the Osawa model), we see that the line obtained from this function is also equal to - Ea / R (in this respect it differs from Osawa). Equation 4.11 expresses Flynn and Wall's model mathematically. b is constant for reactions of order 1 (Saron & Felisberti, 2009).

$$E_a = - (R / b) . [(d \log \beta) / (d1/T)] \quad \text{[Eq. 11]}$$

Among the methods described, the methods used in this study were Osawa/Flynn-Wall, Model Free Kinetics and Starink.

2 - EXPERIMENTAL

2.1 - Material Synthesis

The material was synthesized using the coprecipitation method at variable pH, which consists of adding a saline solution containing the two cations to be introduced into the lamellae in 2:1 equivalence to another solution containing the anion to be intercalated (in this case carbonate) in a 2.0 mol.L^{-1} NaOH solution. The mixture was subjected to a 24-hour hydrothermal bath (Daute et. al, 2002), then the materials were matured for 72 hours and subsequently filtered and the crystals washed to pH=7.0. The HDL system synthesized was Mg-Al-CO3 and Zn-Al-CO3. The material was characterized using XRD analyses, which were processed using a Shimadzu diffractometer, with step size = 0.05 and 2ϴ in the 1.5°-60° range.

The x-ray diffraction data shows that the material of interest was successfully obtained, as illustrated in Figure 3.

Figure 3 - Diffractogram of the material synthesized in this study.

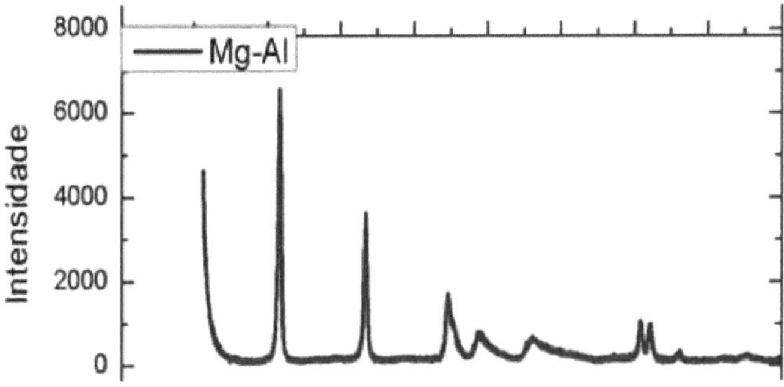

2.2 - Thermogravimetric analysis

The thermogravimetric analysis data obtained in this study was carried out on a Shimadzu© TGA50H apparatus under an atmosphere of N_2, with a flow rate of 50 mL/min, in a temperature range of Tamb-900°C. The heating rates at which the Mg-Al and Zn-Al double lamellar hydroxide samples were subjected were 5, 10, 20 and 25 °C/min. The data analysis was divided according to the stages of metaphase formation: HTD (dehydrated material) and HTB (deoxidized material). Figure 4 illustrates the thermogravimetric curves obtained in this study.

Figure 4 - Thermogravimetric curves for different heating ratios for Mg-Al HDL

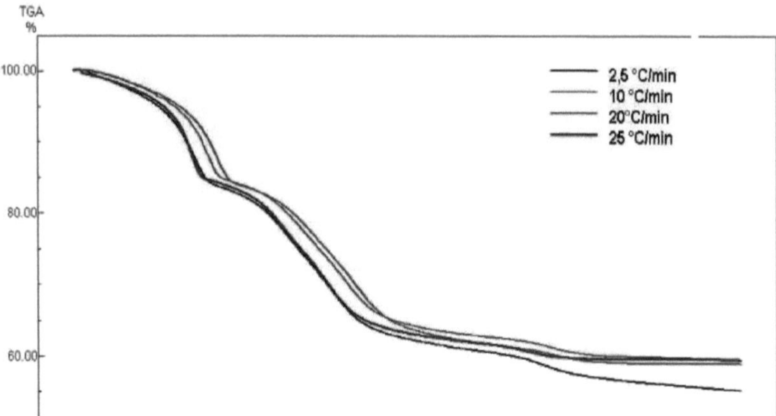

The data from the thermogravimetric curves show that heating at a rate of 2.5 °C/min causes a greater loss of mass in the material sample, as the heating dynamics are slow and the material decomposes in detail. The other samples showed a similar total loss of mass.

2.3- Dehydration of HDL Mg-Al

Figure 5 shows the Osawa/Flynn-Wall graph for the dehydration stage of Mg-Al lamellar double hydroxide. The linearity of the data obtained shows that the method is consistent in describing the kinetics of the thermal dehydration process of the material studied for heating ratios of 5, 10 and 20°C/min, indicating that these processes at these heating ratios present an order of 1° degree. For a heating rate of 25°C/min there is a discrepancy in the linearity of the function, which indicates that the increase in the energy supplied to the sample over a shorter period of time caused a change in the dynamics of the process studied. It is worth noting that these observations do not take into account the conversion rates of 0.1, 0.15 and 0.2, as these conversions are achieved at very low temperatures and are therefore related to extrinsic water, so obviously there is no coherence to the model, as the loss of extrinsic water is a random process.

Figura 5 - Osawa/Flynn-Wall Graph for the Mg-Al HDL Thermal Dehydration Stage

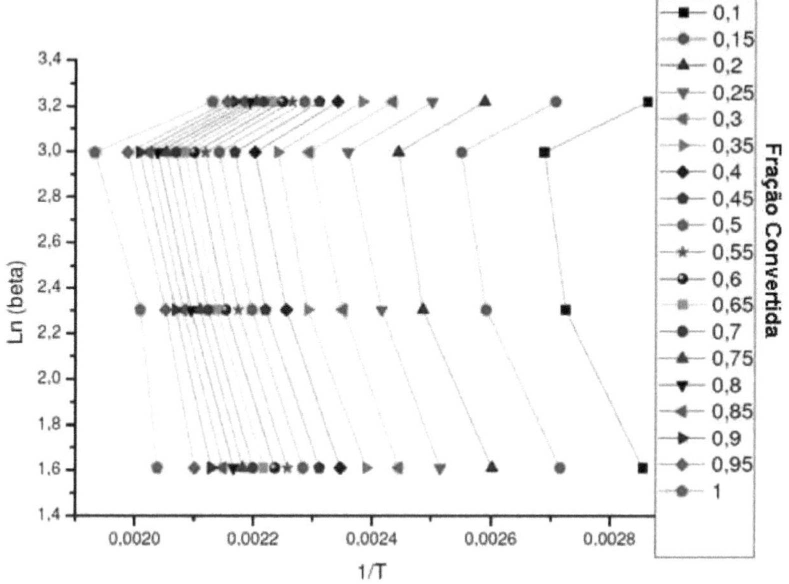

For a conversion rate of 1, there is also a slight deviation from the model, due to the change in the stage of the thermal decomposition process. Immediately after dehydration, deoxidization/decarbonization begins. The loss of the last infinitesimal fraction of water molecules is certainly infinitesimally simultaneous with the start of decarbonization, i.e. the kinetics are altered by the presence of this fact.

A study of the activation energies of the process considering the Flynn and Wall and Osawa models is illustrated in Figure 6. As illustrated, the behavior of the activation energy data is equivalent and coincides with the information contained in the literature regarding the existing stages in the formation of HTD metaphase (Stanimirova et. al, 2004; Forano et. al, 2006; Yang, et. al, 2002). The two ridges are noticeable in the evolution graph of the activation energy for both models. The first ridge, whose conversion rate is equal to 0.15, occurs at relatively low temperatures and is attributed to extrinsic water which is very weakly bound to the surface of the material through Van der Waals forces, which are released at this stage of the process. The second ridge, whose conversion rate is 0.45, is attributed to the beginning of the breakdown of the octamer-bicycle organization that the water molecules acquire due to the interaction between them because of the hydrogen bonds, since the corresponding temperature range concerns the phase transition of water from the liquid to the gaseous state. From this point onwards, the loss of water becomes more pronounced for the same time interval than at lower conversion rates, as they reach a higher entropy state due to the breaking of hydrogen bonds in a relatively high temperature range. However, the literature reports the existence of a certain stage in which some water molecules contained in the interlamellar space that are bound to the brucite-like layers of the lamellar double hydroxide break this interaction. This stage was not identified in the analysis of the evolution of the activation energy using the Osawa and Flynn-Wall theoretical models.

Practically the same behavior of the evolution of the activation energy described for the Osawa and Flynn-Wall models was obtained by the Starink model (illustrated in Figure 7). These models were not expressed in the same graph due to differences in the order of magnitude of the activation energy values.

Figura 6 - Activation Energy of Thermal Dehydration of HDL Mg-Al by Osawa and Flynn-Wall Models

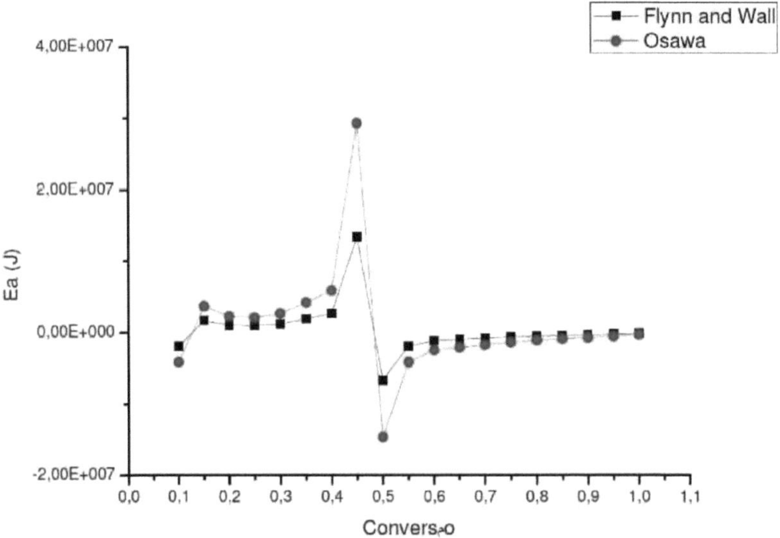

The behavior obtained in the Osawa, Flynn-Wall, Starink and "Model Free Kinetics" models is similar. It is worth noting that there is no variation in the activation energy at the moment of approximately 70% of the process. This crest would be in a conversion range in which the bonds between the water molecules and the brucite-like layers are broken. This indicates that the water molecules present in the middle part of the lamella are released more easily than those present near the brucite-type layers, since, in addition to the interaction between the surrounding water molecules, there is interaction with the layer. Figure 7 shows the activation energies for the Starink and Model Free Kinetics models. However, at such high temperatures there is no noticeable variation in the activation energy, since the phase change itself takes place at much lower temperatures.

Figura 7 - Activation energies for the Starink and Model Free Kinetics models.

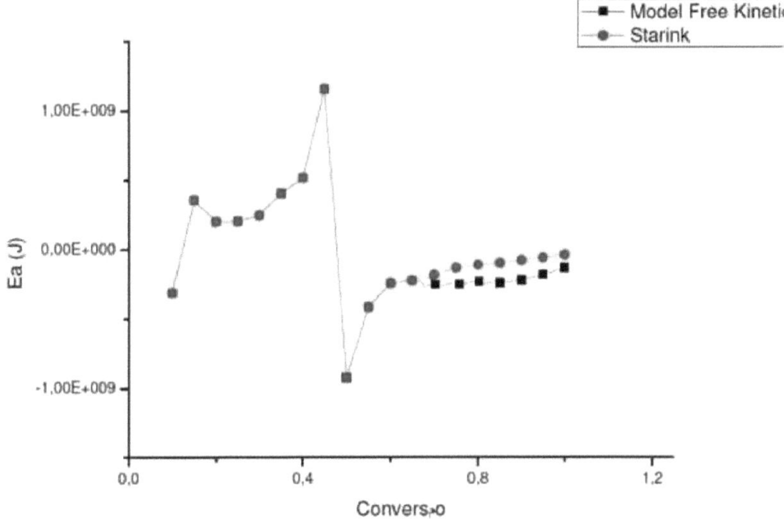

2.4 - HDL Mg-Al deoxidation

According to the specialized literature, deoxidation occurs in three stages, where 1/3 of OH^- is released in each stage (Stanimirova et. al, 2004). As approximate amounts of hydroxyl are released in each stage, one would naturally not expect such significant variations in the activation energy considering the process as a whole. The activation energy graphs from the Osawa and Flynn-Wall models reflect this reality. The deductions of the activation energy according to the Flynn-Wall model considered that the deoxyhydration process is practically linear, only when the deoxyhydration stage reaches the end (95%), i.e. the groups involved have already been completely removed and the remaining groups are easier to remove, even because of the very high temperature, is there any decrease in the activation energy. This behavior was also observed by deduction using the Osawa model. Figure 8 shows the activation energies obtained from the Osawa and Flynn-Wall models.

Figura 8 - Activation energies obtained by the Osawa and Flynn-Wall methods

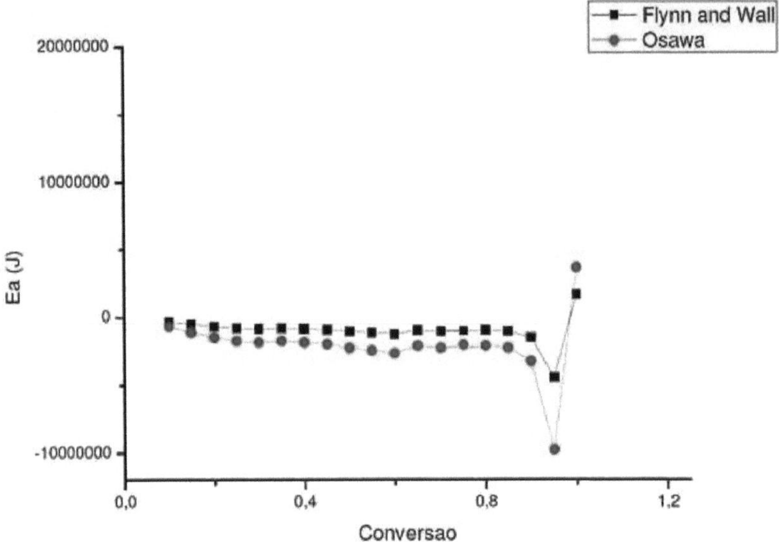

The behavior of the activation energy in the process deduced using Starink's model showed the same behavior as that shown by Osawa's model (including the 95% conversion rate, where there is a decrease in the activation energy).

This is quite natural, since one is derived from the other. The kinetic description by Model Free-Kinetics presented the deoxidization process as continuous, as suggested by Stanimirova and collaborators (Stanimirova, 2004), whose activation energies remained constant at practically all stages of the process. Figure 9 illustrates the activation energies deduced by the Starink and Model Free-Kinetics models. As in the previous situation, these models were not plotted together due to very large differences in the order of magnitude of the activation energy values obtained.

Figura 9 - Activation energies for the Starink and Model Free Kinetics models

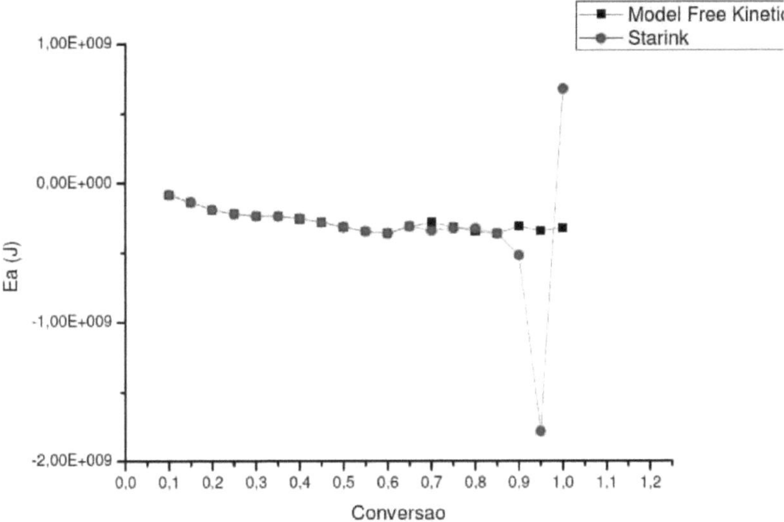

3 - CONCLUSIONS

The kinetic parameters obtained using the Flynn-Wall, Osawa, Starink and "Model Free Kinetics" models clearly confirm the deductions described in the literature using data from combined techniques (Stanimirova et al., 2004; Forano et al., 2006) regarding the formation and decomposition of metaphases during the pyrolysis of a lamellar double hydroxide. This indicates that kinetic models based on thermogravimetry data offer an excellent elucidation of the process without the need to use combined techniques.

These data also reveal that in the formation of the HT-D metaphase, the water molecules strongly bound in the structure of the material are released at a conversion rate of 45%. In the HT-B metaphase, the activation energy is practically constant throughout the process, indicating that there are similar losses at all conversion rates.

REFERENCES

ATKINS P. Physico-chemistry: Fundamentals. 3rd Ed. Sâo Paulo: LTC; 2001.467p. ATKINS P, JONES L. Principios de Quimica: Questionando a vida moderna e o ambiente. 6ª Ed. Sao Paulo: Bookman; 2006. 965p.

Biagini E, Lippi F, Petarca L, Tognotti, L. Devolatilization rate of biomasses and coalbiomass blends: an experimental investigation. *Fuel,* 2002 May; 81(8):1041-1050, doi.org/10.1016/S0016-2361(01)00204-6

BROWN ME. Introduction to thermal analysis techniques and applications. 2nd Ed. New York (USA): Kluwer Academic Publishers, 2004. 267 p.

FLYNN J H, WALL L A. A quick, direct method for the determination of activation energy from thermogravimetric data, Journal of Polymer Science Part B: Polymer Letters, 1966 May:4(5);323-328, DOI:10.1002/pol.1966.110040504

FORANO C, HIBINO T, LEROUX F, TAVIOT-GUÉHO C. Handbook of Clay Science: Developments in Clay Science. Amsterdam (Netherlands): Elsevier Academic Press. 2006. Chapter 13, Layered Double Hydroxides; p. 1021-1095, doi.org/10.1016/S1572-4352(05)01039-1.

Kissinger HE. Variation of peak temperature with heating rate in differential thermal analysis. *Journal of research of the National Bureau of Standards,* 1956 Oct;57(4):217- 221, DOI: 10.6028/jres.057.026

Mackenzie RC. Nomenclature in thermal analysis, part IV. *Thermochimica Acta,* 1979 Jan; 28(1): 1 -6, doi.org/10.1016/0040-6031 (79)87001 -X

Órfao JJM, Figueiredo JL. A simplified method for determination of lignocellulosic materials pyrolysis kinetics from isothermal thermogravimetric experiments. *Thermochimica Acta,* 2001 Jan;380(1):67-78, doi.org/10.1016/S0040- 6031(01)00634-7

Ozawa T. A new method of analyzing thermogravimetric data. *Bulletin of the chemical society of Japan,* 1965 Feb; 38(11):1881-1886, doi.org/10.1246/bcsj.38.1881

Saron C, Felisberti MI. Influence of Colorants on the Thermo-Oxidative Degradation of Polycarbonate. Matèria, 2009 Oct; 14(3):1028-1038, dx.doi.org/10.1590/S1517-70762009000300014.

Stanimirova T, Piperov N, Petrova N, Kirov G. Thermal evolution of Mg-Al-CO3 hydrotalcites. *Clay Minerals.* 2004 Jun;39(2):177-191, doi.org/10.1180/0009855043920129

Starink MJ. A new method for the derivation of activation energies from experiments performed at constant heating rate. *Thermochimica Acta,* 1996 Oct;288(1-2):97-104, doi.org/10.1016/S0040-6031(96)03053-5

Valente J S, Figueras F, Gravelle M, Kumbhar P, Lopez J, Besse JP. Basic properties of the mixed oxides obtained by thermal decomposition of hydrotalcites containing different metallic compositions. *Journal of Catalysis,* 2000 Jan; 189(2):370-381, doi.org/10.1006/jcat.1999.2706

Vyazovkin S, Sbirrazzuoli N. Confidence intervals for the activation energy estimated by few experiments. *Analytica chimica acta,* 1997 Nov;355(2):175-180, doi.org/10.1016/S0003-2670(97)00505-9

Vyazovkin S, Dollimore D. Linear and non-linear procedures in isoconversional computations of the activation energy of nonisothermal reactions in solids. J Chem Inf Comp Sci, 1996 Jan;36: 42-45, DOI: 10.1021⁄ci950062m

White JE, Catallo WJ, Legendre BL. Biomass Pyrolysis Kinetics: A Comparative Critical Review With Relevant Agricultural Residue Case Studies, J. Anal. Appl. Pyrolysis. 2011 Jan;91 (1): 1-33, doi:10.1016/J.Jaap.2011.01.004

DAUTE P, FOELL J, LANGE I, KUEPPER S, WEDL P, KLAMANN, JD. US Patent 6,362,261, 2002 Mar.

Yang W, Kim Y, Liu PK, Sahimi M, Tsotsis TT. A study by in situ techniques of the thermal evolution of the structure of a Mg-Al-CO 3 layered double hydroxide. *Chemical Engineering Science,* 2002 Aug; 57(15): 2945-2953, doi.org/10.1016/S0009- 2509(02)00185-9

CHAPTER 4: REMOVAL OF NITRATE FROM WATER USING LAMELLAR DOUBLE HYDROXIDES

SUMMARY

The growth of agricultural activity, coupled with a lack of education on the proper handling of fertilizer use, has led to concern about the increased concentration of these substances in water sources. To this end, it is necessary to study materials and processes that make it possible to consume this water. In addition, these materials need to be adapted to the current sustainable trend, i.e. they need to offer the capacity for regeneration and reuse. In this context, this work studies the removal of nitrate, a common component in agricultural fertilizers, through lamellar double hydroxides and the subsequent reuse of this material.

1. INTRODUCTION

Nitrate is a notorious chemical component in the environment whose main source of human intake is in food and drinking water. One estimate reveals that 85% of the nitrate ingested by a human being is found in vegetables, however, taking into account the current "fast food generation", it is clear that this fact is utopian, with drinking water being the main vector for ingesting this anionic specimen.

Data from the World Health Organization, WHO, reveal a sharp increase in the concentration of nitrate in surface waters in many countries over the past 30-40 years and that the main reasons for this trend have been the increased use of artificial fertilizers, changes in land use and the disposal of agricultural waste (WHO, 2011).

The average intake indicated by the World Health Organization is 43-131 mg of nitrate per day. However, literature data from the 1980s and 1990s shows that the daily intake by the majority of the world's population is alarmingly higher, with the consumption of water, a vital necessity, being the biggest vector for the ingestion of this species (WHO, 1985b; Bonnell, 1995). Certainly, these concentrations are higher. The current water crisis makes this situation extremely worrying.

In view of the above, it is important to study processes and materials capable of removing nitrates from drinking water. In this sense, this work studies the removal of nitrate anions through adsorption processes with lamellar double hydroxides of the Zn-Al and Mg-Al system, thermally treated under different conditions and the reuse of the material in subsequent processes.

2. EXPERIMENTAL PROCEDURE

2.1 Adsorbent preparation

Samples of anionic clays from two systems M -M -A^{2+3+-y} , whose ratio, M^{2+} / M^{3+} =2:1, were synthesized using the coprecipitation method at varying pH (Daute, et. al, 2002). In this method, a saline solution containing the two cations to be introduced into the lamellae of the material in a 2:1 ratio is slowly introduced into another solution containing the anion to be intercalated (in this case the CO_3 $anion^{-2}$) in a strongly alkaline medium (2.0 mol.L-1 NaOH solution). This mixture was subjected to a hydrothermal bath for 24 hours and then the materials were matured for 72 hours and the crystals washed to pH=7.0. The HDL systems synthesized were Mg-Al-CO3 and Zn-Al-CO3.

The material was characterized using X-ray diffraction analyses, which were processed using a Shimadzu diffractometer, with step size = 0.05 and 2ϴ in the 1.5°-60° range.

The adsorbent, calcined HDL, was subjected to Thermogravimetric Analysis, carried out at the same heating rate, and subsequent derivation of the curve was carried out to roughly identify the current phase at the temperature at which the samples were calcined (470°C).

2.2 Adsorption tests

For the nitrate anion adsorption tests in aqueous solution, the HDL samples were subjected to the calcination process. Only Mg-Al and Zn-Al HDLs were used in this study. Calcination was carried out at heating rates of 2.5°C/mim and 20°C/mim up to 470°C. Once the temperature of interest (470°C) was reached, the samples were preserved in a desiccator to ensure the integrity of the material. Four samples were used for the nitrate anion adsorption tests. Table 1 shows the samples, the treatment carried out on the material and the name given to them.

Table 1 - Adsorbent samples used in this study

Material	Heating ratio in calcination	Name of Reference
HDL Zn-Al	2.5 °C/min	ZnAl2,5
HDL Zn-Al	20 °C/min	ZnAl20
HDL Mg-Al	2.5 °C/min	MgAl2,5
HDL Mg-Al	20 °C/min	MgAl20

The nitrate adsorption kinetics tests on calcined HDLs were carried out at time intervals of 5 min, 15 min, 30 min, 1h, 2h and 4h. These tests were carried out in batches, where the aqueous sample and the adsorbent were subjected to constant agitation at 200 rpm on a shaking table. The solution/adsorbent ratio was 500:1. This amount is proportional to 2g per

1L of solution, whose nitrate concentration was 1000 ppm. The quantification of nitrate in the samples was detected by ultraviolet spectroscopy, using absorbance at 220 nm and 275nnm, according to the method described in Standard Methods of Water Analysis (1999).

3. RESULTS AND DISCUSSION

The diffractograms of the prepared materials are shown in Figure 1, where it can be seen that the desired materials were successfully obtained, with Zn-Al being the most crystalline, as illustrated in Figure 1.

Figura 1 - Diffractograms of the samples synthesized in this study

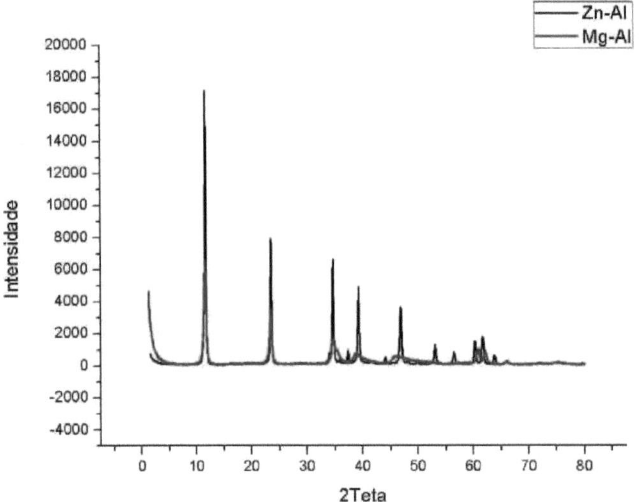

Figure 2 illustrates the TG curves for the HDL samples at the different heating ratios.

Figura 2 - TG curves for HDL samples at different heating ratios

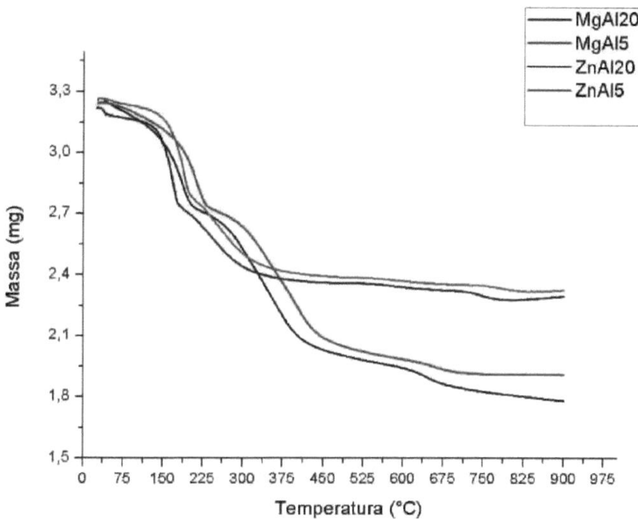

In the temperature dynamics to which the samples were subjected, 2.5°C/min up to 470°C, the material practically acquired the double oxide condition characteristic of lamellar double hydroxides subjected to high temperatures. A precise analysis of the DTG curve showed that the deoxidization of the material in this temperature dynamic occurred at a temperature of approximately 495.62 °C. Between 470°C and 495.62°C there is a mass loss of 0.002%.

The Zn-Al HDL calcined at 2.5°C/min showed that the total loss of hydroxyls present in the structure of the material was reached at approximately 381.41°C. Taking into account the fact that the material was calcined at a temperature of 470°C, it can be seen that the formation of the double oxide metaphase was achieved by the material in this temperature range.

In the Mg-Al HDL sample calcined at a heating rate of 20°C/min, it can be seen that the oxide phase resulting from exposing the material to high temperatures was acquired in a temperature range of approximately 493.03°C. Therefore, the sample used in the nitrate anion adsorption kinetics tests did not completely reach the double oxide phase.

The TG curves for the Zn-Al material, calcined at a heating rate of 20°C/min, show that the double oxide phase was obtained at a temperature of 379.17°C. It is therefore possible to conclude that the sample used as an adsorbent in the nitrate anion adsorption kinetics tests was in the double oxide phase.

Table 2 compares the information contained in the literature (Forano et al., 2006) with the data obtained in this study. The literature data surveyed indicates that the double oxide phase is obtained by subjecting the material to an unspecified heating rate.

Table 2 - Comparison of the temperature ranges used to obtain the double oxide phase of the Zn-Al and Mg-Al lamellar double hydroxides.

Material	Approximate Temperature Range
Mg-Al	~450°C (Forano et. al, 2006)
Zn-Al	~400°C (Forano et. al, 2006)
ZnAl2,5	381,41°C
ZnAl20	379,17°C
MgAl2,5	495,62°C
MgAl20	493,04°C

Comparing the material worked on in this study with the reference material reported in the literature, there is a discrepancy of approximately 10% in the data of interest cited in the table above. These factors are extremely natural and are related to differences in crystallinity, and the experimental conditions have a major influence.

3.1 Adsorption of nitrate ions

Figure 3 graphically illustrates the results of the adsorption tests for the MgAl2.5 sample. As already mentioned, the adsorbent saturation process took place very quickly. With a contact time of 5 minutes, practically the same results were obtained as with the other contact time intervals, while the maximum amount adsorbed with a contact time of 30 minutes was just over 2% less.

Figure 3 - Nitrate removal kinetics of the MgAl2,5 sample

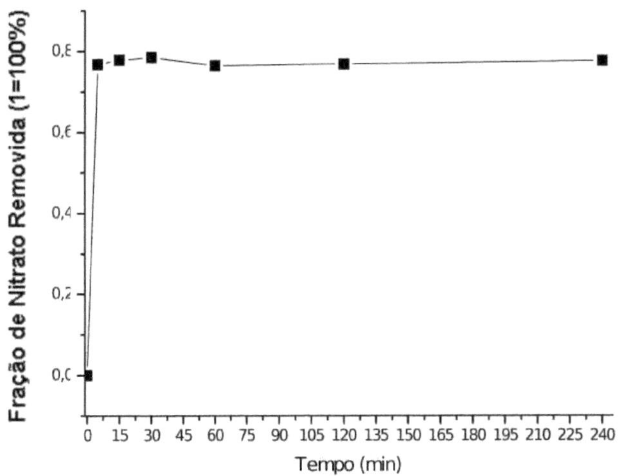

The tests with the ZnAl2.5 adsorbent were carried out in a similar way to the MgAl2.5 sample, i.e. the adsorbent quickly reached its maximum adsorption capacity in a short space of time, so contact time intervals longer than 4 hours were not tested either.

Figure 4 - Nitrate removal kinetics by the ZnAl2,5 adsorbent

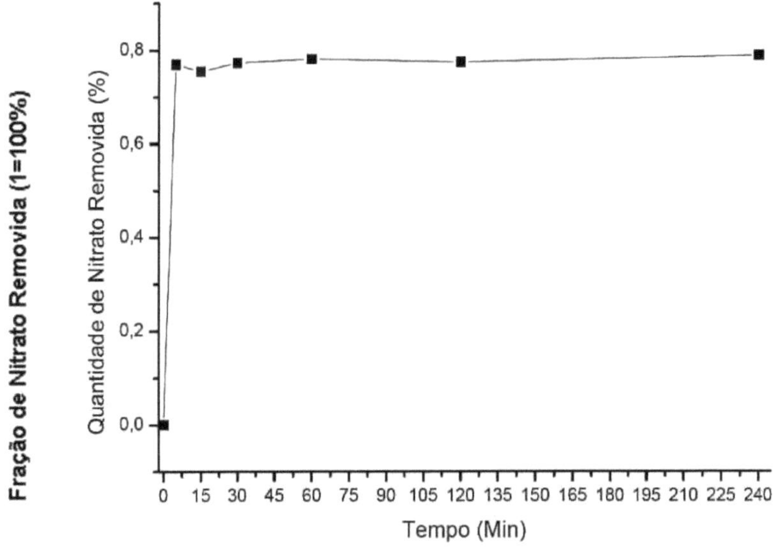

Figure 5 compares the adsorption results obtained by the samples calcined under the same heating dynamics.

Figure 5 - Comparison between samples calcined under the same conditions

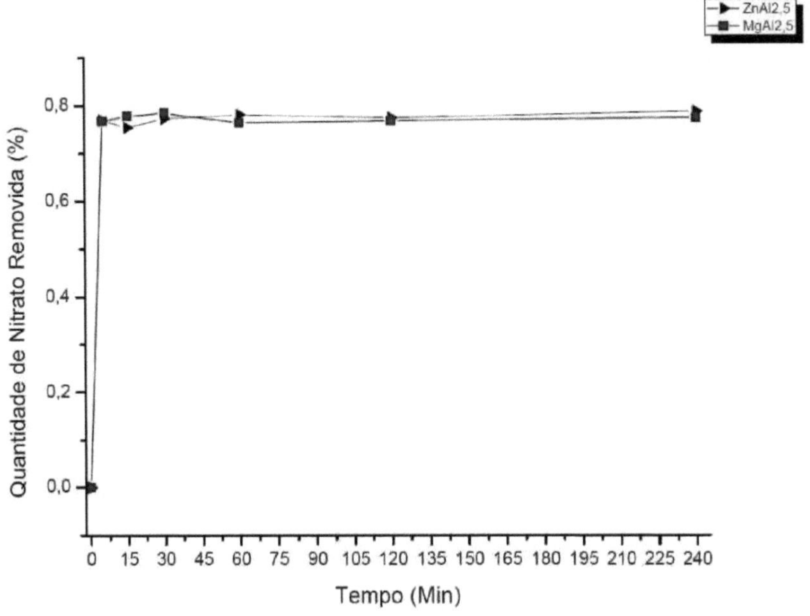

The comparison illustrated in Figure 5 shows that the MgAl2.5 material presented more satisfactory results than ZnAl2.5. In 5 minutes of contact, practically the same amount is removed. At 15 minutes and 30 minutes, MgAl2.5 has a higher adsorption potential.

For the samples calcined at a dynamic temperature of 20°C/min, kinetic curves are observed with some particularities that differ from those of the samples calcined at 2.5°C/min. As explained in Figure 2 and also in Table 2, the samples calcined at 20°C/min reach the oxide phase at lower temperatures than those calcined at 2.5°C/min, so the adsorbents used had a purer oxide phase when calcined at 20°C/min, with the presence of more adsorptive sites due to the loss of the charge-stabilizing anion from the precursor HDL.

Figure 6 illustrates the adsorption kinetics curve of the MgAl20 and ZnAl20 samples. As illustrated, the samples calcined at 20 °C/min showed a speed comparable to the samples calcined at 2.5 °C/min, with similar efficiency in a short time. short time interval. However, unlike the samples calcined at 2.5 °C/min, which adsorbed nitrate quickly and soon reached saturation, the MgAl20 and ZnAl20 samples reached saturation in longer contact time intervals, showing almost 10% higher nitrate removal at the saturation point.

Figure 6 - Nitrate removal kinetics of the MgAl20 and ZnAl20 samples

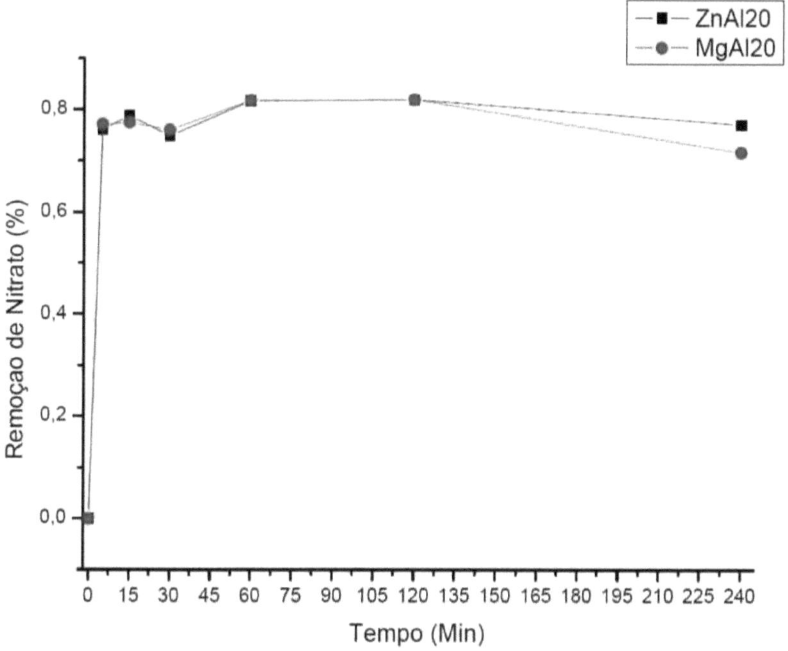

Figure 7 shows a comparison of the kinetic curves of nitrate removal by the samples used in this study. As can be seen, all the samples showed similar adsorption dynamics within a contact time of 5 minutes, with rapid adsorption amounting to 76-77% nitrate removal in this short space of time. As previously mentioned, the samples calcined at 20°C/min showed greater adsorption power, which, however, was obtained at longer time intervals. Thus, we have a parallel in terms of the adsorbent materials, i.e. neither the composition nor the calcination dynamics had any influence on the instantaneous adsorption kinetics, since all the adsorbent materials, considering the shortest time interval studied, showed practically the same efficiency. However, the maximum adsorption power is influenced by the calcination dynamics of the materials, since the materials calcined at a heating rate of 20°C/min have an adsorption power whose superiority is close to 10% in efficiency, but this potential is reached at high time intervals, considering the fact that >70% of nitrate is removed in just 5 minutes of contact.

Figure 7 - Nitrate removal kinetics of the samples used in this study

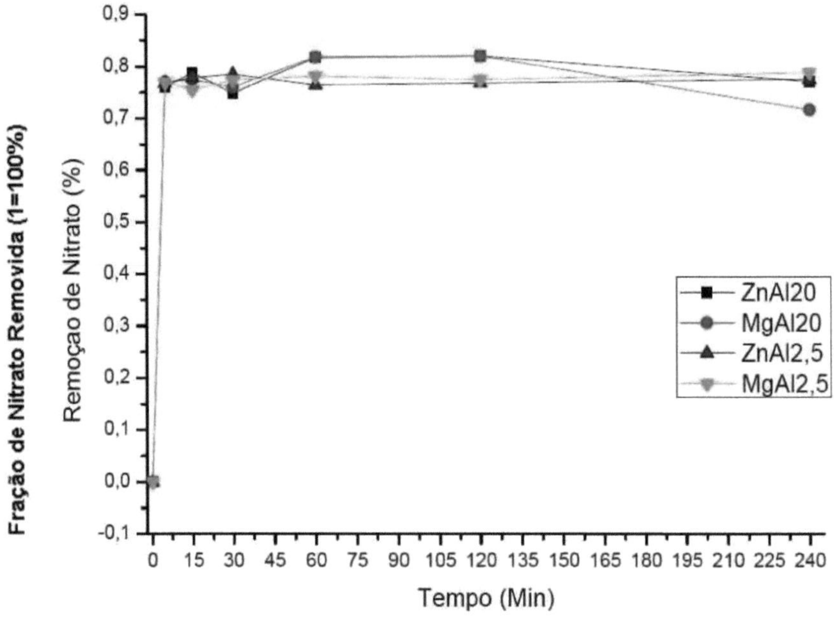

Table 3 shows the data obtained from the experiments carried out in this study.

Table 3 - Nitrate Removal Data from Calcined HDLs under Different Conditions.

Time (Min)	Removal MgAl 20 (%)	Removal ZnAl20 (%)	Removal MgAl2,5 (%)	Removal ZnAl2,5 (%)
0	0	0	0	0
5	77,07	76,0	76,79	77,0
15	77,48	78,8	77,81	75,0
30	76,02	74,83	78,48	77,0
60	81,8	81,7	76,43	78,0
120	82,0	82,0	76,82	77,0
240	71,7	77,1	77,56	79,0

3.2 Regeneration and Application of the Adsorbent in Successive Processes

In order to demonstrate that the adsorbent can be reused in successive processes, samples of HDL Zn-Al were subjected to the regeneration process, which consists of calcining the material at 20°C/min, followed by the nitrate adsorption test.

Figure 8 shows the data obtained from this study

The data illustrated in Figure 8 shows that the material did not lose its adsorptive capacity during the adsorption-regeneration-adsorption processes. These results indicate that the adsorbent maintained a similar potential in all the stages undertaken.

Figure 8 - Nitrate removal by the regenerated adsorbent

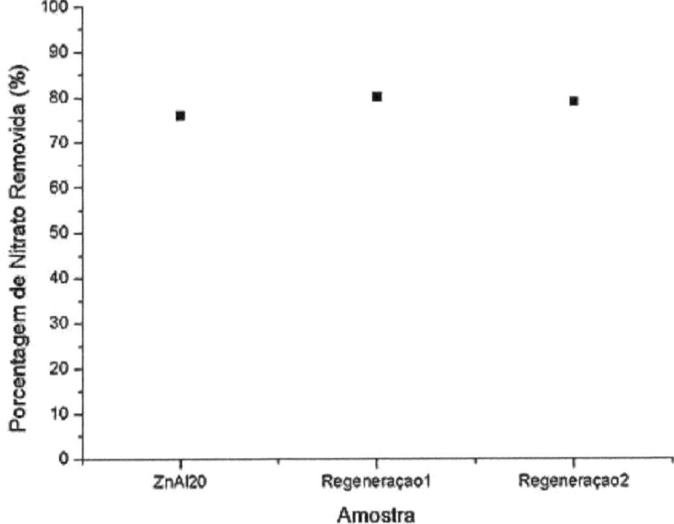

4.FINAL CONSIDERATIONS

The application of a contaminant-removing material must take into account its potential in a real situation, i.e. a material that has a high nitrate-removing power does not always do so under conditions suitable for sanitary and industrial applications, since a material that takes a long time to remove a considerable amount of contaminant would cause losses in terms of cost and benefit. In this case, it is worth noting that the adsorbent has reached its potential in an extremely short time, making it very promising for its purpose. In addition, there is the possibility of recovering and reusing the adsorbent in subsequent applications, which increases the relevance of the potential of lamellar double hydroxides for the application illustrated.

However, among the materials used in this study, the materials calcined at a heating rate of 20°C/min showed better adsorptive power than those calcined at 2.5°C/min. However, these results were obtained with a contact time of 1 hour. At lower contact time intervals, the samples showed practically the same efficiency as nitrate anion adsorbents.

Tests using the regenerated adsorbent in subsequent processes show how promising the applicability of the respective material is, since even in sequential applications the material did not lose its adsorptive capacity in the short term.

REFERENCES

AMERICAN PUBLIC HEALTH ASSOCIATION et al. Standard methods for the examination of water and wastewater. American Public Health Association, 1999

Bonell, A. E., Nitrate concentrations in vegetables. In: Proceedings of the International Workshop on Health Aspects of Nitrates and its Metabolites (Particularly Nitrite), Bilthoven: Council of Europe Press, 1994, p. 11-20.

Daute, P., Foell, J.; Lange, I., Kuepper, S, Wedl, P., Klamann, J.d. US Patent 6,362,261, 2002.

FORANO, Claude et al. .1 layered double hydroxides. Developments in clay science, v. 1, p. 1021-1095, 2006.

JOINT, F. A. O. et al. Guidelines for the study of dietary intakes of chemical contaminants. 1985

WHO, Geneva. Guidelines for drinking-water quality. World Health Organization, v. 216, p. 303-4, 2011.

FSC
www.fsc.org
MIX
Papier aus verantwortungsvollen Quellen
Paper from responsible sources
FSC® C105338

Printed by Books on Demand GmbH, Norderstedt / Germany